DO CATS HAVE ESP?

DO
CATS HAVE
ESP?

JEANE DIXON

RUNNING PRESS
PHILADELPHIA · LONDON

AARON PUBLISHING GROUP, INC.
NEW YORK

The publishers express their gratitude to Norma Langley
for helping to bring this project to fruition.

PUBLISHED IN 1998 BY
Aaron Publishing Group
685 Third Avenue
26th Floor
New York, New York 10017

Printed in U.S.A.

Illustrated by Rustyn L. Birch

9 8 7 6 5 4 3 2 1

Library of Congress Cataloging-in-Publication Number 98-67029

ISBN 0-9665202-0-3

This book may be ordered by mail from the distributor.
Please include $2.50 for postage and handling.
But try your bookstore first!

DISTRIBUTED BY
Running Press Book Publishers
125 South Twenty-second Street
Philadelphia, Pennsylvania 19103-4399

CONTENTS

DO
CATS HAVE
ESP?

SINCE I am a psychic as well as a cat lover, people often ask me whether cats have extrasensory perception. Anyone who has lived with a cat and who "talks" to cats knows the answer is an unqualified "Yes!"

Although some people claim that cats have only normal animal instinct, I believe that anyone who has spent quality time with a cat knows that its animal instinct is just the beginning of the story. Throughout history, people from all walks of life have recognized the mystic qualities that set cats apart from other animals.

While instinct is a formidable component of the cat's survival skills and overall personality, true extrasensory perception (ESP) gives some cats the ability to predict danger and an uncanny awareness of the moods, health, and well-being of their human companions. It is extrasensory perception that makes it possible, too, for some cats to perform truly remarkable feats of navigation, a unique feline gift. Stories of homing cats seem to indicate that the greatest odds of time and distance cannot keep a cat from finding a "lost" master.

From my own experience with feline friends throughout my lifetime, I've collected stories of hero cats and homing cats and other remarkable examples of feline prowess. Many people feel as I do about cats and some have documented their findings. Extrasensory perception in cats is an accepted concept around the world and has been since the time that cats were first domesticated. History and folklore are full of stories that bear witness to the extraordinary powers of these household pets.

An acute sensitivity to human feelings was one of the traits that first made cats prized as pets. There is evidence of people relying on the healing power of cats from as far back as ancient Egypt. Many other ancient cultures also believed that cats not only warned of the presence of illness, but could take onto themselves some of the pain suffered by humans.

Like most people who own cats, I have been comforted by my cat who has spent endless hours at my side, soothing me with total love and devotion through common colds and the most life-threatening of illnesses.

Traditional science now seems to be catching up with the instincts of cat lovers. Recent studies have shown that having a pet extends a person's life span and enhances the quality of life—especially for those people who live alone.

But some cats go much farther than others. Some cats actually save lives and there are many stories of cats who have successfully monitored the well-being of their masters.

Ringo, a tabby from Bowling Green, Ohio, for example, recently became only the tenth cat in the twentieth century to be honored by the American Humane Association with the coveted Stillman Award for heroism.

Ringo's owners, Carol and Ray Steiner, couldn't understand why both of them were suffering the same symptoms, including high blood pressure, memory loss, nausea, and pounding headaches. It was their golden tabby, Ringo, who ferreted out the trouble and was able to communicate to Carol the cause of the baffling and lingering illness.

One day, Ringo seemed to go crazy—slamming himself against the back door and meowing until Carol followed him outside to the gas meter in the yard.

"Ringo began digging in the rocks—and cats hate to dig," Carol reported. When she leaned over to see what was in the hole, she was nearly overcome by the fumes of natural gas. Later, a representative of the gas company told her the gas was at explosive levels. Not only was there a danger of explosion, but the gas was continually seeping into the house, where the Steiners were slowly being poisoned to death.

"We probably would have gone to sleep one night and never woken up if it wasn't for Ringo's incredible intelligence," Ray said.

Cats seem especially attuned to the well-being of babies. Abigail, for example, was an eight-month-old Persian devoted to her master's two-year-old son, Christopher. One day, while his mother was gardening outside, Christopher was sleeping inside the house. He woke from his nap and began to choke.

Sensing the baby's danger, Abigail came to full alert. The cat scratched and cried and howled at the screen door until Christopher's mother came to the rescue. If it weren't for Abigail's raising of the alarm, the child would have choked to death.

Kandy, a California mother, had a similar experience. While she was doing the dishes one day, her usually quiet and well-behaved tabby, Oscar, began to act strangely. The cat set up a howl and jumped up onto the drainboard.

After repeated attempts to quiet Oscar and shoo him away, getting her heels nipped in the process, Kandy followed the agitated cat as he ran back to her baby's bedroom. She wanted to catch him and put him outside before he woke Anthony, her four-month-old son. She was astonished to see Oscar jump right into Anthony's crib.

Immediately Kandy forgot the cat and picked up her breathless child, who had obviously spit up and was choking on the vomit. Kandy knew she had to act fast, so she tapped Anthony between the shoulder blades and the baby coughed up the obstruction, took a deep breath, and howled.

After the crisis, Kandy realized that Oscar broke all the house rules to get her attention—and to save her baby's life.

Bernita Rogers and her husband, Roy, of Fort Leavenworth, Kansas, had suffered through three unsuccessful pregnancies before Bernita gave birth to their daughter, Stacey, in 1986. Although premature, Stacey was a small, but healthy, baby. However, when the five-pound infant came down with a cold six weeks after leaving the hospital, Bernita was naturally concerned and rushed the child to the doctor. Thinking the new mother was overreacting, the doctor advised that she put a humidifier in the baby's room.

Bernita brought Stacey home and put her down for a nap, and then tried to relax in another room. But Midnight, the family's cat, refused to leave Bernita alone—he wouldn't stop jumping on her lap and pawing around her legs. "This wasn't his usual behavior," Bernita said.

Although Bernita was successful in shooing Midnight away, the cat persisted in trying to get her attention. Suddenly she heard the cat through the baby monitor, emitting an eerie, moaning sound from the nursery. When Bernita reached Stacey's room, she found a frantic cat perched atop the bureau, and her baby daughter blue and gasping for air.

Stacey was rushed to the hospital, where she was resuscitated and diagnosed with a viral infection. Had it not been for

Midnight's persistent warnings, Stacey surely would have died.

There are many cases where cats seem to make conscious decisions to sacrifice their own lives for their masters.

Spinner is a good example. His master was ill and apparently fell asleep smoking in bed. When the man's wife returned from shopping, she smelled what she thought were burning feathers. In fact, it was Spinner's fur that was on fire.

The cat had discovered a fire in the mattress and must have realized that the smoke was dangerous. He probably tried to rouse his master, but the man had taken medication that made him sleep soundly.

So Spinner covered the smoldering hole in the mattress with his body, not budging until his master's wife entered the room. He allowed himself to be burned rather than let the fire spread to his master.

The left side of Spinner's body was badly burned, but the hero's wounds healed and he lived to be sixteen years old. Today, the family still cherishes his memory.

Throughout history, many stories have been recounted of extraordinary cats. One of my favorite tales about a cat protecting its master comes from fifteenth-century England.

Richard III, an English king with a reputation for cruelty, had Sir Henry Wyatt, a leader of the opposition Lancasterian party, locked in the Tower of London and left to starve. Sir Wyatt's cat found him and every day slipped through the bars with a freshly killed pigeon, which was then cooked for him by a kindly guard. Wyatt outlived the king and his family returned to high status, never forgetting their debt to a cat.

Some cats display an amazing ability for taking care of people. Tabitha is such a cat, and an animal training school in Tucson, Arizona, is her legacy.

In 1974, Mardi Hatfield suffered head injuries in an automobile accident. She was faced with a future fraught with potential danger: she had frequent seizures and could fall down and lose consciousness anywhere—in the middle of a busy highway or at the top of the cellar stairs.

Early into her medical nightmare, Mardi realized that her cat, Tabitha, was giving her warning signals before the seizures occurred. Somehow the cat was aware of the impending episode and would jump up and down and pull on Mardi's clothing.

Although the seizures continued, Tabitha's warnings enabled Mardi to prepare for them by sitting or lying in a safe place. Once Tabitha saved Mardi from having a seizure at a dangerous highway crossing. "Without Tabitha, I could have been in the middle of a busy street and lost control of myself. I could have been killed."

Now Mardi never leaves home without a cat on a leash. Through Tabitha, Mardi learned that cats can sense potentially deadly seizures, and she began training other animals to aid disabled people. She's trained, for example, a hearing-ear cat for a deaf person. That cat signals when the teakettle whistles and when the doorbell rings. Another of Mardi's cats serves as a "nurse's aide" in a retirement home. He patrols the halls, signaling a nurse when a patient is in distress.

Many cat lovers swear that from the actions of their cats they learn of far off family trouble—sometimes even the deaths of family members.

17

At four o'clock one morning in 1990, the Naguski family was asleep in their Pennsylvania apartment when their cat, Goliath, jumped off a rocking chair and began to race around and literally bounce off the walls.

A few hours later, Janice Naguski received a phone call and was told that her father, a great favorite of Goliath's, had had a heart attack during the night and had died.

In a letter published in the February 1991 issue of *Cats* magazine, Janice wrote, "Goliath and my dad were buddies . . . I believe before passing beyond this world that my dad stopped to say good-bye and that Goliath could sense his presence in the room."

Janice added, "I would like my experience to be shared . . . so that those who have had similar experiences will know they are not alone."

Cats will sometimes demonstrate their extrasensory perception by throwing caution to the wind in order to respond to the call of a beloved master.

Carol Ann Timmel not only talked to her cat, she established a response code so that when she called, Tabitha was to come on the run.

(This is a training rule that I wholeheartedly endorse. Always establish a special call—not the cat's name—that serves as an instant response code. That way, if it is in danger, caught in traffic, perhaps, the cat will not override its defense patterns and run to you if you call its name as a warning.)

When Carol Ann moved from New York to Los Angeles in 1994, she bought a cargo crate in which Tabitha and her sister, Pandora, would travel on the plane. When Carol Ann went to the cargo area to pick up the cats, Tabitha was missing. The plane was searched, but she was not to be found.

Tower Airlines conducted numerous searches and left tuna for bait, but after twelve days airline personnel were ready to call off the search. During that time, the plane had traveled thirty thousand miles on its regular air route and no one had seen or heard a cat.

Carol Ann finally persuaded the airline to ground the plane and allow her on board to search. A psychic (perhaps in some way "reached" by Tabitha) had pinpointed the dropped ceiling of the cargo hold as the place Carol Ann would find Tabitha.

Hearing Carol Ann's special call, Tabitha poked her nose out from overhead and gave a relieved "meow."

Nearly two weeks without food or water took two pounds off the striped tabby's normal eight-pound weight, and Carol Ann vowed that she would keep her cats grounded in the future.

A recent fire in New York City made an unnamed feline jumper famous. I like to think of him as the flying tiger.

Onlookers watching fire spurt from an apartment building were dismayed to see a black-and-white tabby trapped on a fifth floor balcony. Then their dismay turned to amazement as the cat leaped into thin air and landed on its feet on the street in front of a little girl.

A fireman at the scene reported: "The little girl picked him up when he landed and said: 'He's mine.' It was amazing to watch." The fireman added, "There was no way he should have survived. I hope his other eight lives run smoother."

It is well documented that cats can find their way home—or reunite themselves with masters from whom they've been separated—against seemingly impossible odds. Scientists call this phenomenon psi trailing and it may be the most baffling extrasensory ability demonstrated by cats. Many cases have been reported of cats who followed families for uncharted miles rather than change masters. Often the psi-trailing cats leave perfectly good homes, arranged to let them live out their lives in familiar surroundings, to reunite with the one they love.

There seem to be two kinds of homing instincts natural to cats. The first is the ability to find their way home from far away. This is demonstrated by cats who have been kidnapped or given away. The second type of homing is the remarkable ability some cats have to find people who have moved to places the cats have never been.

Stories of such homing cats may strike doubters as implausible, but they hold little surprise for cat lovers.

Beau Chat, for example, vanished from his home in Louisiana while his young master, Butchie, and his family were away looking for a new home in Texarkana, Texas. Returning to pack up their old home, the family found the cat missing and searched everywhere without success. Finally, on moving day, the broken-hearted boy had to give up the search.

Five months later, Beau Chat turned up at Butchie's new elementary school. Butchie knew his cat immediately, and his parents were convinced, too, because of a tar smear on the Persian's tail and a distinctive scar over one of its eyes. But it was the family dog, Alexander, who made the final judgment, when he allowed Beau Chat to sleep curled up with him—a privilege never accorded another pet.

Beau Chat had traveled almost three hundred miles to find

Butchie. With no map, no help, and never having been to the town of Texarkana, the cat had found its master.

Sometimes pets are the target of thieves who underestimate the power of an animal's love for its master.

Brenda James picked a poor time to visit her brother in Manchester, England. His house was burglarized during her stay and Miss James's red tabby, Cindy, was taken as part of the loot. Nearly eighteen months after Miss James returned to her home in London, Cindy found her mistress even though Miss James had moved to a new apartment while Cindy was away. Manchester to London: two hundred miles.

There are times when circumstances force us to leave our cats to the care of others. But even when we put them in a loving home, we can't be sure that we won't some day be reunited.

When a nurse from Sandusky, Ohio, took a job in Orlando, Florida, she gave her cat, Li-Ping, to her sister. The black tom couldn't get comfortable in his new home and within a fortnight he disappeared.

A month passed before the nurse was confronted with her loving pet. Sitting on the porch of her new Florida home, the nurse called to a bedraggled black cat that was making his way down her street.

The cat responded to his name and leaped into the nurse's arms. Although Li-Ping was in terrible shape, the nurse was able to make a positive identification because her cat had an abnormal throat condition, and did not make the usual cat sounds. He could only communicate with a unique rasping noise, which the happy nurse had learned to love. This cat who found his mistress

21

had traveled more than fifteen hundred miles from Ohio to Florida.

Then there's the case of Murka of Moscow, the cat who returned from exile. Her story was reported in 1989 by *Pravda*, the then leading newspaper in the Soviet Union.

Vladimir Dontsov had two pets: Murka, a gray-and-white tabby, and a canary. But when the canary accidentally escaped its cage, the cat killed the bird. Dontsov forgave Murka, bought another canary, and forgot the incident. But when he found that Murka had taught herself to open the birdcage door and helped herself to a second canary dinner, Dontsov banished the cat to Voronezh to live with his grandmother.

Murka stayed for two days, then started the four-hundred-mile trek back to Moscow, alone and on foot. Besides adventure, *Pravda* reported that Murka also found romance. She who left in disgrace, came home a pregnant heroine.

Some cats set speed records in finding their way home. In 1949, a cat named Rusty traveled more than one hundred miles a day to catch up with its master, who had moved the almost one thousand miles from Boston, Massachusetts, to Chicago, Illinois. A British tabby, McCavity, walked five hundred miles in three weeks to get back to its old home in England. But sometimes it takes a cat months or even years to find its master.

One of the best documented stories of a cat finding its master comes from a veterinarian who had left his cat with his staff at a New York animal hospital when he moved to California—a door-to-door distance of more than two thousand miles.

Five months later, the animal surgeon returned to his apartment to find a cat curled up asleep in a chair—the same nesting place of the cat he had in New York. The cat looked the same, acted the same, and the doctor reported that he felt the extrasensory pull of his familiar pet. But even better than that were his medical notes: if this was the same cat, an X-ray would show an enlarged bone at the fourth caudal vertebra—and, yes, it was there.

When the Smiths sold their California home to retire to Florida, they believed their aging cat, Tom, would be happiest if he were left in the house where he had been reared.

But Tom languished and the instinct to find the Smiths stirred. He was made of sterner stuff than they anticipated. Although he had the run of his old home, his masters were gone—and soon, so was Tom.

More than two years later, when a mangy cat showed up twenty-five hundred miles away in their new backyard, Mrs. Smith was justifiably shaken. Mr. Smith came to the rescue, but was overwhelmed when the "strange" cat jumped into his arms and began to lick his face.

It seemed to Mr. Smith that Tom had somehow crossed the boundaries of time and space to travel from California to Florida in a courageous two-year odyssey to find his family.

At first, Mrs. Smith didn't accept the incredible possibility that this was their cat and that he had found them without any help. After all, Tom didn't even know the direction they took when they left California. But Pablum was the test that convinced her. Tom had learned to love it as a kitten, and when he buried his whiskers in the cereal bowl that Mrs. Smith put in front of him, she knew that it was really her Tom.

Humans are not the only beneficiaries of the love and often life-giving devotion of cats. Nursing queens have been known to willingly accept orphaned puppies, raccoons, squirrels, even baby rats, into their litters and suckle them as if they were their own.

Few mothers have gone to the extraordinary lengths to save the lives of their children as did Scarlett, the cat who became world famous in March 1996 for the miraculous rescue of her five kittens.

New York City firefighters had finished battling a fire in an abandoned Brooklyn building when they were surprised to hear a chorus of meowing. There, on the street, sat three crying kittens. The firefighters soon found the mother cat across the street, where she had managed to carry two other kittens to safety. Badly burned, her strength was finally spent after searching out and rescuing all five kittens, one by one, from the blaze.

Although her eyes were swollen shut by her burns, when the firefighters reunited the cat with her entire brood, she touched each of her kittens with her nose to make sure they were all safe.

Scarlett, as she was later named, and her four surviving kittens were brought to the North Shore Animal League in Port Washington, New York, which received over seven thousand letters about Scarlett, twelve hundred offering adoption. The heroine cat has become a symbol for abandoned animals, and even has her own World Wide Web page, "Scarlett's Web," as part of the NSAL site on the Internet. The shelter has established The Scarlett Fund to raise funds for injured and abandoned animals.

The site features this poem, written by Rosemary Asmussen, in tribute to a mother's love.

FROM A HEROINE

Why is everyone so surprised
That I saved my furry five,
That in spite of pain and danger
I brought them out alive?
True, my eyes were barely open
But I heard their frantic wails;
Through smoke and flames I saw
Scorched ears and burning tails.
Every trip was a burdened choice
But I could make no other.
The rescuers have called me cat
But I am also "mother."

THE
PSYCHIC
CAT OF
MYTHS

THANK goodness we are living in an enlightened age when a psychic can admit a fondness for cats. At various times in history, neither was safe if they hung around together.

Since cats were domesticated more than three thousand years ago, people of every culture and age have been aware that cats have extrasensory perception. And a look at times past makes one thing clear: Cats are such extraordinary creatures that entire cultures have either revered or feared them. From the time of ancient Egypt, when cats were worshipped as gods, to the Middle Ages, when cats were so reviled they were almost driven to extinction in Europe, to our modern coexistence with cats, these animals have had a remarkable journey.

The ancient Egyptians first domesticated the wild cats of Africa and prized them for their ability to keep the nation's grain stores free of rodents. Called Myeo, or Mau, the cat's status in Egyptian society grew, and eventually became the subject of worship, taking the form of the goddess Bastet, who had the body of a woman and the head of a cat. A daughter of the goddess Isis, the protector of the dead and the great magician whose power transcended that of all other deities, Bastet was the goddess of fertility and of the sun and the moon. She was said to control the life-giving heat of the sun, because it was believed that a cat's eyes, which appear to glow at night, held the light of the sun captive.

Killing a cat in ancient Egypt was a crime punishable by death. A Persian king, embroiled in a war with Egypt, ordered his troops to gather up all the cats they could find. He then armed his front lines with the live cats and won the battle when the Egyptian army surrendered rather than kill the cats borne as shields by the Persian soldiers.

It was also illegal to take a cat outside the country's border. The cat was so deeply associated with fertility that it was believed the removal of cats from Egypt would leave women barren and bring an end to their civilization.

Nevertheless, many traders smuggled cats out of Egypt to other Mediterranean countries, and evidence of worship of Bastet has been found in ancient Rome and Pompeii. Bastet's influence can be found farther north in Europe as Freyja, the Scandanavian goddess of fertility, who is often depicted riding into battle in a chariot drawn by cats.

In Siam, now Thailand, cats lived in temples and palaces, and it was believed that the souls of departed people lived in the bodies of sacred cats before passing on to the next life.

In Burma, this same belief gives rise to a wonderful story that tells of the creation of the extremely attractive breed of cat known as the Birman. Legend has it that the golden goddess with sapphire-colored eyes, Tsuyn-Kyan-Kse, presided over the transmigration of souls. She resided at the Temple of Lao-Tsun with one hundred white cats and many priests in attendance. During an attack on the temple, the aged priest Mun-Ha was killed. Upon his death, Mun-Ha's companion cat, Sinh, immediately jumped upon the old priest's body and faced the goddess. It is said that at this moment, the priest's soul entered the body of the cat: its hair turned golden and its legs turned brown, except for the four white feet that rested on the priest's body. The eyes of the cat became the same bright blue as those of the goddess. The next day it was discovered that all the other white cats in the temple had undergone the same transformation. From that time on they were considered sacred.

Although cats are not mentioned at all in the Bible, according to a legend in the Middle East, Noah extracted the first pet cat from the pair of lions on board the ark. Something had to be done about the proliferation of the original pair of rats.

Like Adam and Eve of the Bible, the progenitors of many peoples were believed to have possessed special powers and gifts that were variously lost by later generations.

One of the most magnificent Creation stories comes from the ancient Mayans who inhabited Central America. According to their legends, the first human beings were four individuals called motherfathers. Three of them were named for the magnificent big cat of the region, the jaguar. They were: Jaguar Cedar, Jaguar Night, and Dark Jaguar. When mates were created for the first humans, Dark Jaguar and his wife were the only pair given the gift of clairvoyance.

Ancient Chinese believed that cats brought good fortune and the glow from their eyes would frighten away evil spirits. According to Japanese folklore, the rare, male tortoiseshell is the world's most psychic cat. At the turn of the century, these cats were highly prized on sailing ships because they were believed to have the ability to ward off storms and sea monsters, and to distract shipboard ghosts from interfering with navigation.

All breeds of cats were prized by European sailors of old, because they believed that cats "knew" the way home, and would always reside on the side of the ship closest to port. Wives of

30

British sailors kept black cats to ensure their husbands safe return from the sea.

Unfortunately, the Middle Ages in Europe were truly the dark ages for cats. Leaders of the Christian church began a campaign against cats. Black cats were especially feared, particularly after Pope Gregory IX declared in 1233 that heretics worshipped the Devil in the form of a black male cat. Men and women were tortured or even killed for helping a sick or injured cat. Cats were slaughtered en masse all over England and the Continent, which led to their near extinction by the year 1400.

Why were cats persecuted during Europe's Middle Ages? These were times of great human distress. Something had to be the cause. Cats were associated with evil, witchcraft, voodoo, and black magic. Cats and "witches" were easier to blame than kings who waged wars, or the spread of disease from poor sanitation. Often the scapegoats were the most downtrodden: the mentally ill, the poor, and, most frequently, women without family protection. Many such women hired themselves out as servants and care givers, with the most skilled in the healing arts becoming midwives. The combination of skills beyond those of others of their station, and the reliance on their pet cats for company, made these unprotected women of England and of America's New England the target of witch-hunts. Cats suffered along with their mistresses.

I believe the cat was chosen for persecution because it exemplified freedom. Cats travel at night, and their aloofness and independent nature may have added to the belief that they were evil. Therefore, the killing of the "haughty" animal was a cruel example, used to keep the people in line.

While the cat population dropped, rats and disease-carrying fleas were left to proliferate. Ridding Europe of cats left the people exposed to one plague after another. More than a quarter of Europe's population died from bubonic plague in the three years from 1347 through 1350.

Both dominant religions of Europe martyred cats. The Tudor Queen Mary (1553–1558) burned cats as a warning to Protestants. Elizabeth I (1558–1603) burned a cat-filled effigy of the Pope at her coronation as a warning to Catholics. That practice was repeated annually as a tribute to the queen.

Fortunately, some countries, like Norway, never succumbed to witch-hunts. They never stopped loving their useful, hard-working cats and kept the European breeds from absolute extinction. In pockets of Europe, including monasteries, rational officials looked the other way when saucers of milk were set on doorsteps. Thus western breeds survived in small numbers as companions, as well as defenders of the grain.

This terrible period of persecution lasted about four hundred years, but never spread to India or the Middle and Far East. By the eighteenth century, cats were once again looked upon favorably. In 1835, a law was passed in England forbidding the mistreatment of any animals.

Children of many cultures have been given cats to teach them important qualities. Victorian girls were given cats so they could study the dedication of motherhood; boys the concentration of the hunt, which would be helpful in making their fortunes. In many cultures to this day children are encouraged to observe cats as an example of wisdom.

Respect for cats in the Middle East is embodied by the story of the prophet Mohammed's love for his cat, Muezza. It is said that rather than disturb Muezza while she slept on his cloak, Mohammed cut the sleeve from his garment. Legend also credits Mohammed with giving a blessing to Muezza that ensured the cat's safety—always landing on its feet.

In this enlightened age, we know our cats to be loving friends, neither gods nor devils. But it is true that there is more to cats than many people, those who are not cat fanciers, comprehend. The cat, with its two-sided nature—haughty but sweet, loner but nurturer—continues to fascinate and baffle humankind.

Someone has claimed good or bad luck for every movement a cat makes. A yawning cat, for example, means opportunity is coming. A cat who rubs against your legs ensures good luck. If a cat looks at you right after it washes, you will be the first in the group to marry.

Black cats got a bad name from Pope Gregory IX, but in the East it is the white cat that is associated with bad luck. Still another legend has it that on every black cat there is a single hair that is white. If you can remove it without the cat scratching you, this white hair will bring you wealth or luck in love.

It seems that with the exception of our own culture, most peoples believed the black cat was the lucky one. Buddhists, for example, respected all cats of all colors, but the home of a dark cat was a magnet for gold, while a light-colored cat attracted silver. In England, if a black cat crosses the path of a couple as they leave their wedding ceremony, it predicts a long, happy union.

Similarly, some believe that if a cat sneezes near the bride on the morning of her wedding, she will have a happy marriage. And in southern France, people say that if a young, unmarried girl steps on a cat's tail by accident, she will meet her future husband twelve months from that day.

From Japan comes the charming story of the "beckoning cat," or the *maneki neko*. It is told that there was once an old shopkeeper who had lost all of his customers. With one bowl of rice left and no hope of earning money to buy more, the shopkeeper was about to begin his last meal when a small white kitten entered his store. Thinking that his life was at its end, while the kitten's was just beginning, the shopkeeper gave his bowl of rice to the little cat and went to bed. The next morning, the shopkeeper awoke to find his store filled with customers and many more people outside waiting to come in.

"I don't understand what has happened," the shopkeeper said to a customer. "Just yesterday I thought that all was lost."

"It is the little white cat that sits in your front window," the customer replied. "There is not a person who can resist when it raises its paw and beckons one to come in and shop."

To this day, ceramic figurines of cats with one paw raised are sold throughout Japan. It is said that, when placed facing the front door of a business establishment, the beckoning cat will bring good fortune.

The legends of many countries offer stories of so-called "magic" cats, which will bring wealth into the house where they are well fed. The English children's story of Dick Whittington's cat, which brought its master great fortune, is one such tale. In France, such a cat must be seduced with a fat chicken, and the future owner of the cat must carry it home without looking backward. If the cat is given the first bite of every meal, it will give its master a gold coin every morning.

One of the many myths to arise from the association of cats with witches is the belief that a cat has nine lives. It was said that a witch could change into a cat nine times before her powers left her.

Shape-shifting, or morphing, is a universal theme of literature. Many cultures have myths that revolve around humans transforming themselves into cats, and cats transforming themselves into humans. It is no mystery why we would like to claim the gracefulness of the cat as our own.

Much of our fascination with cats surrounds their magnificent and mysterious eyes. In the dark, they seem to hover alone, disembodied and shining brightly on the darkest of nights. Ancient peoples believed the cat had captured a piece of the sun which it called up at will to see in the dark. And the cat's habit of looking people straight in the eye without blinking or flinching keeps alive the idea that cats can read the minds and emotions of humans.

So the cat's all-seeing eye became the most obvious link between a psychic and a cat. One who can claim clairvoyance, or clear sight, "sees" beyond the confines of space and time. Modern clairvoyants know their gift is all too human and all too limited. We have come a long way from attributing superhuman powers to our cats as the ancients did.

I believe cats "see" and feel many things that humans, like myself, cannot. I believe in the extrasensory perception of cats, but I believe clairvoyance is a gift that is reserved for human beings.

THE
PSI
CAT

SCIENCE doesn't just happen. It is the compounding of knowledge. Some of today's most important scientific studies in animal behavior are based on centuries spent observing cats. Just as medical researchers analyze the healing plants of folklore, animal behaviorists gather insights from myths.

Myths are based on inspiration and creativity as well as observations by many people over many years. While kernels of truth may be buried under tons of superstition in mythology, it is surprising how much sound information ancient peoples were able to record about the animals around them.

The ancients believed, for example, that even ferocious, wild animals sometimes helped "good people." According to legend, a wolf reared Romulus and his twin brother, Remus, who later founded the city of Rome. Was an early Roman slave named Androclese really spared by a lion who recalled the Christian's removal of a thorn from its paw as a cub? Edgar Rice Burroughs's Tarzan stories are based on African legends, still told on that continent, of a human child raised by gorillas.

Certainly everything is not yet known or understood about the depths of empathy between people and animals. Without question, however, animals do communicate with people and scientists investigating animal intelligence no longer question whether animals think and communicate. Decades of experimentation with whales, dolphins, monkeys, parrots, and cats have brought science to the next question, "How do animals think?"

Back in the 1950s one remarkable scientist was a pioneer with his study of the questions raised specifically by cat myths—Do cats have extrasensory perception?

The conviction that cats have extrasensory powers, widely held over time, led Dr. Joseph Banks Rhine, head of the Parapsychology Laboratory—now called the Rhine Institute—at Duke University in Durham, North Carolina, to study extrasensory perception in cats as well as in human beings.

Along with other distinguished work involving mind reading, out-of-body travel, and clairvoyance in humans, Dr. Rhine's staff put the phenomenon of ESP in humans and cats, as well as the mystery of homing cats, to a variety of tests. Dr. Rhine coined the phrase "psi-cat"—"psi" being a word that loosely approximates "psychic."

Dr. Rhine summed up his philosophy of animal ESP in an article "The Present Outlook on the Question of Psi in Animals" that was published in the *Journal of Parapsychology.*

In the article he wrote, "There is clearly a lot to explain in animal behavior that could be due to ESP and is not yet otherwise accounted for. We have even reached a second stage, that at which we find behavior that can be accounted for by nothing else that is known and experimentally verified, except extrasensory perception. . . . It is the only known principle that is adequate."

Recently my associate, Norma Langley, spoke to Dr. Karlis Osis, an animal behaviorist, now retired, who worked with Dr. Rhine in the Parapsychology Laboratory at Duke University.

Most of the cat studies at Duke were completed under scientific laboratory discipline. In addition to clinical tests, the lab set out to study cats under "normal" conditions. In that pursuit, Dr. Osis, his wife, Carla, and their ten-year-old daughter, Gunta, took cats into their home to bond with the family.

The main focus of Dr. Rhine's studies of extrasensory perception involved telepathy. According to Dr. Osis, "We worked with cats because people have always felt there was telepathy between themselves and cats. And we found significant evidence that there is telepathy between humans and cats. . .but not all humans, of course. Our best results came from working with kittens who interacted best with researchers who liked cats."

Dr. Osis worked on a series of thought-transfer experiments using his daughter as the controller. The child would put food in two dishes of equal size and then try to steer each cat telepathically to the left or the right. Taking into consideration random selection, in a significant number of trials the child was able to direct the actions of several cats.

At the same time his cat experiments were being conducted, other researchers at the Parapsychology Laboratory were continuing Dr. Rhine's study of human extrasensory perception. Highly psychic people were being put through batteries of tests and rated for their ability to distinguish forms displayed in distant sites, to communicate their thoughts to others and to read the thoughts of others, and to project themselves to far-off places through out-of-body travel. Some of these people were also tested with cats. According to Dr. Osis, the more psychic the individual, the better he or she was able to control the activities of the cats.

The most amazing incidence of mind-bonding between a psi cat and a human occurred when Olaf Jensen, the famous Swedish psychic, met Chippy. Dr. Osis described the experiment: "We had a nice big, black Persian named Chippy. He wouldn't let anyone strange near him. But he took immediately to Jensen. They seemed to be talking to each other with their minds.

"Jensen would tell us where Chippy would go and what he would do next, and then he would 'talk' to Chippy—who sprinted off to do his bidding. They responded to each other as ESP friends."

Another Duke experiment that demonstrated the mental bonding of cats and people was accomplished by a psychology student who claimed to be able to travel out of body at will.

The student was tested with his seven-week-old kitten. First it was established that the cat was peaceful when the student was in the room with her, and she would pick up his mental directions. When the student left the cat alone in a laboratory room, it became agitated and noisy.

The research team blocked out the floor of a room into a grid, so they could measure the young cat's activities when the student was physically present, when the student was absent, and when the student was physically absent, but was projecting himself into the room through out-of-body travel (OBT).

In repeated tests, researchers established that the kitten was not only agitated and moving all the time the student was away, but also meowed an average of four times per second. When the student returned to the room the cat quieted. Vital signs, noise, and activity all subsided. On occasion the kitten would curl up around the student's feet and take a nap.

After the pattern of actual absence and proximity was established, the second part of the experiment took place: the student left the cat, went into a distant cubicle, and tried to quiet the agitated animal through projection, using ESP.

In a significant number of instances, the student was able to calm the cat as well through OBT as when he was actually in the same room. Using the grid to show the cat's movement in the cubicle, the student was also able to move his cat from corner to corner by giving psychic OBT directions.

The experiment concluded that the kitten became agitated when left alone, and was calmed by the student's physical presence as well as by the student's OBT presence.

At the peak of their experiments, Dr. Rhine and his associates worked with about thirty-six cats at a time. Dr. Osis recalls, "Sometimes people would bring us cats they felt were extraordinary. But most of our cats were given to us by people who simply had too many."

There have always been stories about cats who could find their way home despite the greatest odds. These tales of psi-trailing inspired another area of investigation at Duke University.

The researchers at the Parapsychology Laboratory sent out a call for "homing cats." They made it clear that all stories would be documented by researchers. In addition, the animals had to be available for examination, had to have distinctive markings, and the owners as well as other eyewitnesses had to attest to the cats' journeys.

At the same time, Dr. Osis tried his best to get his own house cats lost. "We started on a small scale," Dr. Osis recalled. "Our house was in the woods and, like the Pied Piper, I would parade my cats through the trees for about one-third of a mile and then leave them. They had no trouble finding their way home from there.

"When we were sure a cat considered our house its home, we devised some stringent tests. To 'lose' a cat, I would drive it through town to a far-off meadow, where I would leave it.

"We had one haughty cat we called Princess. She obviously considered herself the queen of our household and all the other cats in it. She always found her way home.

"Then we tried a new experiment. Another researcher transported her to a field far from the house. She was placed in a cage. To say she was insulted was an understatement. She complained all the way, and she was one angry cat when she was released in the tall grass. My daughter and I were well hidden before the experiment began.

"When Princess walked out of the cage she turned and looked right at me. I expected her to march directly to our hiding place. Instead she did an about-face and marched in the opposite direction into the woods. She never came back out. We never saw her again. You have to be careful of a cat's feelings.

"One of my favorite stories about the homing instinct," recalled Dr. Osis, "involved a big male cat who was given to us by people who had too many. I kept the cat in my house and my daughter and I thought he was happy there. But after a week, when I put him outside for the first time, he went straight in the direction of his old home. He traveled about ten miles through the outskirts of Durham, North Carolina. When his original owners called to report that he was home, I went and got him and brought him back in my car, taking a circuitous route.

"We tried the same experiment again one week later. Again

he marched straight home. That time, his owners decided to keep him."

Here are the stories of some of the true psi cats investigated by the Parapsychology Laboratory.

One was Clementine, who followed her family from Dunkirk, New York, to Denver, Colorado. Waiting until after she delivered and weaned a litter, she walked sixteen hundred miles through unknown country in four months. Clementine was identified by seven toes on her front paws, and white spots on her stomach and near her left shoulder.

Tom was another impressive cat. He had always lived in Kokomo, Indiana, and then he was shipped by train to Augusta, Georgia. He stayed for three weeks, then set out for Kokomo, covering thirty miles a day on his seven-hundred-and-twenty-mile journey. It took Tom twenty-one days to walk home.

Sugar, a cream-colored, part-Persian cat with a deformed left hip joint disappeared when his family moved from California to Oklahoma. Instinct led him back to his old home. But when Sugar realized the family was gone for good, he managed to find them in Gage, Oklahoma. Sugar leaped through an open downstairs window and onto the shoulder of his mistress. It had taken him fourteen months to travel fifteen hundred miles.

Smokey of Tulsa, Oklahoma, had unusual chin markings. And when his fourteen-year-old mistress played the piano, he would sit next to her, always on her right, with his paws on the keys. When the family moved to Memphis, Tennessee, he disappeared. His instincts led him back to his old home in Tulsa, but he stayed only two weeks before joining the family in Memphis.

Sampson was lost while his family was taking a holiday in Wales, more than two hundred and fifty miles from his home in London. It took him two years to find his way home.

Stories of psi cats of all ages continue to be documented around the world. A news report from Pisa, Italy, tells of an eight-month-old kitten sent by a family to relatives three hundred miles away. The kitten was banished because it insisted on sharing the crib with the family's baby. The kitten not only found its way back home, but hopped directly into the crib when it returned.

Although work at the Parapsychology Laboratory at Duke documented many scientific phenomena, there were problems that researchers could not resolve. Unanswered still is the question: Is our intelligence a help or a hindrance to extrasensory perception?

Dr. Osis stated that after all of his work with animals, he is not sure whether brain size has a great deal to do with extrasensory perception. He found even small-brained chickens showed some trace of extrasensory perception. "The size of the brain seems to be a factor, but not always," he said. "Mice have smaller brains than chickens, but they exhibit a greater depth of extrasensory perception.

"My biggest regret," concluded Dr. Osis, "is that no cat in the experiment showed the slightest interest in sending a telepathic message back to the scientists.

"There are plenty of situations where cats need their

extrasensory perception. But they don't need to use much to socialize with humans. We seem to want to set up telepathic channels more than cats do.

"Of course, they could have been dangling us like puppets and we simply did not catch on. Cats pick the times they are willing to communicate. Sometimes you feel like you have had a lover's quarrel with your cat and you have no idea what you have done to offend it. Other times a cat will let you know exactly what it thinks of you. Then, when you are lucky, and it is in the mood, a cat will display great affection.

"Maybe we moved too quickly or tried to bond with too many cats at once. Maybe it simply is not possible to duplicate in a lab the bond that grows between a cat and a human being in a home setting."

Although Dr. Rhine often had to deal with disbelievers when he published his studies on extrasensory perception, he understood the basis of the skepticism. He wrote, "Human beings have a hard time accepting psychic and other unusual phenomenon because it is so 'improbable.'"

What was improbable in the 1950s, however, is getting closer to the probable as the millennium approaches.

Nutty, for example, a one-year-old, ginger-colored tom cat from Surrey, England, was recently given a Golden Arthur Award. Nutty received his prize—a trophy and the sterling equivalent of approximately sixteen hundred dollars—for his ability to forge a link with his young master, Simon Grossmith. Simon is a five-year-old who suffers from autism and no human has been able to break through the barriers of his affliction. Nutty has happily endured being carried around in a backpack and even being bathed as part of his special relationship with the child.

Other members of the animal kingdom continue to exhibit extraordinary communicative powers. It has now been demonstrated, for example, that chimpanzees, the most studied animals because of their similarities to humans, reach the intelligence level of three-year-old children. At Georgia State University, a chimpanzee named Kanzi can respond correctly to simple requests like "Put the vacuum outside." Chimps at Ohio State University can count objects and then indicate the correct number by selecting a printed card.

Orangutans from Borneo and Sumatra and great apes from Africa have demonstrated three different communication abilities: They can follow spoken instructions, they can use and understand sign language, and they can be taught to count. The National Zoo in Washington, D.C. has a "think tank" where orangutans are learning to communicate with visitors through computer images.

A parrot named Alex at the University of Arizona, Tucson, could identify the color, shape, and material of hundreds of small toys and verbalize his answers.

Researchers at the Pavlov Institute in Russia continue to study animal parapsychology. One of their theories proposes that extrasensory perception may be the result of microscopic rods in the eye that send out and receive telepathic signals like radio waves. At the same time, a researcher at the Ukrainian Institute of Psychology believes "The brain radiates a special, hitherto unknown type of energy responsible for telepathy."

My Most Important Chat with a Cat

EVERY cat lover "talks" to his or her cat. From the cat's point of view, these first conversations take some effort, although according to experts, a talkative cat has about a hundred sounds in its vocabulary.

Although cats respond to our need to communicate with them, they don't often verbalize with other cats. They use their double set of vocal cords in only a few situations, including crooning in courtship, a queen calling her newborn kittens, kittens identifying themselves to their mother, screaming to scare opponents in a catfight, or when in extreme danger.

Beyond its mysterious qualities, a cat is the most real, down-to-earth, honest animal you will ever meet. When it is hungry, it lets you know. But once you have fed it, it will leave and bother you no more—until it is hungry again. A cat is not a flatterer, currying favor, but it knows how you feel about it.

I think that through their extrasensory perception, cats can communicate with anyone who loves them. To try communicating with your cat, sit down with it in a quiet place. Talk to it as you would to another person. The secret of making it understand you is the tone of your voice. If you speak with genuine goodness in your heart, your cat will respond, and you, in turn, will sense its answer.

Until a kitten figures out what you want, it may seem to ignore your invitation to converse. Keep talking to your new or young cat; soon it will catch on. Friendship takes time. You have to have the patience and wisdom of a cat and be able to free your mind of distracting things when you sit beside it. Yet, as this is going on—as you become aware of the cat's sincerity, patience, and its acceptance of reality—you will be learning a great deal from your pet.

Although cats can't repeat our words because of differences in throat structure, they instinctively pick up the rhythm and patterns of our speech—and will try to echo it. In time they learn to "talk" with meows, murmurs, purrs, calls, and cries, as well as with their silent extrasensory perception.

After you have learned to distinguish the meanings behind the range of sounds your pet makes, you have the key to understanding other talkative cats, too. Since I've had at least one feline housemate most of my life, and because I have always been curious to find out what a cat knows that I don't, I frequently walk up to a strange cat and exchange pleasantries.

I was very grateful that I was in good practice with "cat talk" when my work took me to Japan a few years ago. It was there that I had my most important chat with a cat.

Psychics live unusual lives. Although I no longer do private readings, there are still old friends, some of them highly placed in government and business, who occasionally question me about future trends. It was an invitation to give some advice on such a matter that led me to Tokyo, where a beautiful Turkish Van helped me break the language barrier.

This is a story I have never revealed before. I was invited to Japan by a group of businessmen who were worried about the activities of a government agency that was key to international trade agreements. The committee, as I will call them, had sought the advice of economists, accountants, and their own most highly respected Japanese psychic. They didn't want to believe what they heard, but all clues pointed to a brewing scandal.

They first contacted me by letter, asking for a reading, which I respectfully declined. Within a week, a phone call came to me from a friend, a Japanese woman from a well-known business family. She interceded on the committee's behalf and I could not turn her down.

The assignment began with a simple request for a "psychic meditation." A gentleman's birth statistics were supplied to help me "channel in" on his future. I did not want to know the man's name or his job—in fact, I didn't want to get involved at all, for I knew little about the intricate, unwritten rules of conduct that govern men of honor in Japan.

After my report was written and delivered, my psychic barometer told me I was at the beginning of a long and stormy journey.

Indeed, within a month, three Japanese men appeared at my office and began a campaign to persuade me to go to Tokyo.

"No, thank you," I said. "I'm too busy to travel. Please, no gifts. Yes, this is my final decision."

The gift they brought was a beautifully carved wooden box that opened to reveal a miniature shrine made of mother-of-pearl. "No strings attached," they assured me. To refuse it, I feared, would be disrespectful. My husband, Jimmy, and I agreed that going to Japan was out of the question. "No time," said I. "You'd be gone too long," said he. With force of will, I pushed back the nagging voice of foresight that was laughing at us both. I was not to be persuaded!

We said cordial farewells, and as the men were leaving they promised Jimmy that they would return the next day. "How long will you be in Washington?" Jimmy asked. "However long it takes," they answered.

A week later, Jimmy gave up. It was not so much what the gentlemen said on their visits to our office, but their polite, unwavering dedication that convinced my husband these men were on a mission. And so he encouraged me to go with them to deliver my message in person.

By the end of the week, I had my reservations booked for a trip to Japan. The young woman who first contacted me on the matter and her husband were to act as my guides and interpreters.

Japan was a magical place. One of the things that surprised me most were the cats—not all of them were oriental breeds. In fact, most that I saw were English tabbies. At the temple GoTo-Ku-Ji, I saw prayer boards filled with the names of lost cats—and each day the monks prayed for their benefit.

After a day to recover from jet lag and another of sight-seeing, it was time to meet with the committee and expand on my reading of the future of the Japanese official.

A limousine had been sent, but as I got in my escorts were diverted to a second car, leaving me alone with an English-speaking driver who immediately pulled away from the curb and took me to an office building.

Another escort was waiting at the door of an open elevator. My friends were nowhere to be seen. The turn of a key sent the elevator to the highest floor, and I entered a room decorated in the simple opulence that one would expect to find in a private home. The room was a large rectangle; its windows painted with a beautiful view.

I was surprised to see about a dozen gentlemen in business suits seated around a low table that held nothing but a silver flask of water in front of the cushion on which I was invited to sit.

There were no clipboards, pens, or recording devices. No one was going to take notes. They were simply ready to grill me politely but unmercifully.

You see, my initial reading had concluded that the revered gentleman in question would eventually be forced to resign from his government post, even though he was completely innocent of wrongdoing. His only alternative was to retire before the trouble began.

Although at the time I did not understand the reason, I was later told that it is customary for the head of an organization or government department to take full responsibility for any trouble that occurs within the ranks. And if the trouble is great enough, the boss normally resigns.

The room was silent after I spoke. This was serious business and everyone was uncomfortable. I was a foreigner, a stranger, and therefore an unknown quantity. The matter at hand was delicate, since it foretold the need for change that would cause a ripple effect through many organizations and many families.

I had been separated from my friends so that no word of the meeting would leave the confines of the conference room. The gentlemen at the table took no chances that information of our meeting would be leaked to the press. I would have liked to have given good news, but I could not suggest another course of action: the official must resign or he would lose face.

I was rapidly losing my audience. In a moment, I felt many would leave, and both my host and I would be embarrassed.

At that moment a door opened and a woman entered the room, followed by a cat. She sat at the only open place at the table and smiled at me. Her cat walked over to me and said hello.

The cat was breathtakingly beautiful. The movement of its elegant body was completely fluid, and its silky white hair had no visible markings. Most remarkable of all was its eyes—one was green and the other was blue. I was later able to identify its breed as Turkish Van.

I spoke to the cat the way one does when meeting an attractive stranger. "Hello," I said. "Where did you come from? You are very beautiful. Can you understand me?"

People adept at talking to cats know instinctively to keep their sentences short and their tone of voice gentle. After you speak to a cat, then you must listen. A talkative cat will jump right into the conversation—usually before you finish with your questions.

When encountering an unfamiliar cat, I speak in a soothing voice. At the same time, I concentrate on the cat's vibrations. This gets the cat's attention. Eventually a low-level psychic connection can be made with almost any friendly cat. How quickly the cat becomes attuned to me and the depth of its understanding depends largely on the cat's intelligence. In the case of the Van, a highly intelligent cat among an intelligent breed, a meeting of minds occurred unusually quickly.

I had forgotten that the cat in my lap was accustomed to hearing Japanese. As I waited expectantly for answers, the cat looked at me with similar questions written all over its face. "Who are you? Why do you speak so strangely?" it asked.

In this case the foreign language and my unfamiliar scent confused the cat at first. So I spoke again: "Tell me, what is it? Who are you?"

The cat answered by licking my hand.

"Thank you," I said out loud, while sensing the full attention of the disturbed committee. To the cat, I whispered, "Can you get me out of here?"

With that, the cat began to emit a series of sounds that included chirrups and gentle mews. Her message was clear. The cat told me that the men in the room were as fearful of the outcome as I. I looked up at them and knew that the cat was right!

Many at the table were wide-eyed at my conversation with the cat. Others looked confused. Some began to smile when they saw I was smiling at them. At least I, among the company, was now much more relaxed after my welcome from the cat.

The man who was sitting farthest from me actually looked relieved. It turned out he was the cat's master and trusted his cat's judgment of people. In fact, it was he who had summoned the silent woman and the chatty cat.

My relief was short-lived. My host suggested that since I was here in Japan—and the other psychics and advisors had concluded much the same outcome—I should meet with the official and tell him there was trouble in his department.

The same day, I met with the official, although the cat did not accompany me. To make a long story short, an international incident was avoided thanks to the interception of a very wise and talkative cat.

MIKE
THE
MAGICAT

THE most remarkable cat I've ever had was Mike, the Magicat. It was Mike who got me started talking to people at Duke University about cats and extrasensory perception. Dr. Joseph Rhine, the chief of Duke's Parapsychology Laboratory, was interested in Mike as a homing cat—until he found out that the White House, where Mike regularly went, was less than three miles from home. In 1967, soon after the Duke University investigations ended, I wrote a magazine article about Mike's ESP.

I've never purchased a cat. When I was a child, there was always a number of barn cats sure to produce enough kittens for a big family like ours. We'd inspect new litters as they were born, often bringing a special kitten into the house as a personal pet. So I began talking to cats and recognizing their ESP potential before I discovered not everyone believes that cats have extrasensory perception.

When my husband, Jimmy, and I set up our first Washington, D.C. home, it was in a small apartment. We decided to wait until we moved into a proper house to have pets. Knowing how I missed having a cat, Jimmy had a secret plan to surprise me with a fancy pedigreed cat as soon as we were settled. Jimmy never got the chance to buy a cat. Mike appeared within days after we moved into our new home.

Actually, Mike came to us quite mysteriously. A well-dressed woman rang our doorbell. We were expecting a guest so Jimmy and I went to the door together. Luckily, I got there first.

A woman was standing there. She told me she had considered drowning her kitten, but instead, she reached down into the huge pocket of her coat and gently brought out this tiny creature.

"Here you are," she said briskly, and walked off. Jimmy, who caught the end of the exchange, started to push past me to chase the woman, but I caught his arm. "She's right, you know," I told him. "This is my cat!"

I handed the kitten to Jimmy, who was muttering and fuming about "the nerve of some people. . . . This is the ugliest cat I've ever seen!" But his gentle hands betrayed his heart—the charisma of that "ugly cat" was already at work. He was a black-and-white, short-haired tabby—in other words, he was an all-American cat!

By the time the first saucer of milk was poured, we were ready to baptize the kitten with a name. Mike is an all-American name, and the nickname of one of our best friends. The season was Christmas. We were thinking of the three wise men who came bearing gifts when we named our kitten Mike, the Magicat.

The story of the gifts of the magi is very close to my heart. They were learned men of their day who heard the prophets foretelling the birth of a special child. The prophets said the time of the event would be marked by the appearance of a great star. The magi had knowledge of astronomy and followed the star to find the child. It was the child to whom they paid tribute; not the clues gleaned from the stars or from prophets.

That neatly sums up my philosophy, too. I read the clues written in the stars, listen to prophets, but worship only one God.

The orphaned cat was my best Christmas present. Looking into Mike's newly opened eyes, I knew we were going to be very special friends. I saw our friendship going on for many years, knowing he would stay with me as long as life permitted. From the first, I sensed an extrasensory connection to my little companion. He was brimming with ESP!

As soon as he grew a little, Mike let me know he wasn't satisfied with being a lap cat. He was too full of energy to hang around the house waiting for me all day, yet he always seemed to know when I'd be there. If he wasn't home to greet me, he'd show up within minutes of my arrival, hurtling through the house until he crashed into my legs.

Even as a kitten, Mike, my "magi" cat helped me through long workdays. Through most of my life, I have gotten up at dawn to start the day with a meditation at sunrise. Then I would go to the James L. Dixon Realty office, where Jimmy and I made our living. As often as I could, I'd come back to the house early in the afternoon to work on "other business." Deadlines for newspapers are as relentless now as they were when Mike was a kitten.

Mike had sensitivity in greater abundance than most cats. Shortly after he came to live with us and I became aware of his magic personality, he began to draw my attention to himself and then helped me concentrate on my own thoughts so that it was easy to shut out distracting noises and sights.

Almost from the beginning, Mike injected himself into my thoughts when he was near me. He seemed to meditate with me. He did not distract me at all, but shared his tranquillity with me.

I often saw Mike just sitting there, as if suspended in a moment of time, and his eyes took on a faraway look. I knew that he was meditating. And when this happened and I had the time, I sat and meditated, too.

While I was out at work, Mike visited the neighbors. My

house was next door to a famous old bookshop and across the street there was a great restaurant. It was a neighborhood where a charming cat could be king.

Mike had his own door to come and go—and go he did, treating all of Washington's Dupont Circle as his domain. I didn't know how wide an area he considered "his" neighborhood until one day, when Mike was about six months old, he disappeared.

Jimmy and I were frantic. We searched everywhere we could think of—but there was no sign of him. His absence was confusing. Psychically, I sensed that he was well and happy, but he certainly wasn't anywhere to be found. Sensing that he might be caught inside someone's house, I advertised for him in a local morning paper. I had his picture inserted in the ad so no one could mistake him.

Three days later, a telephone call came. A woman's voice announced, "This is the White House. We've got your cat." My secretary took the message. When I returned, I called immediately, but no one at the Executive Mansion knew anything at all about my little Mike.

I would not give up, though. After many calls, I found that a secretary—the only one who might have been our mysterious caller—had just left on her vacation. I went to bed that night downcast—but with my faith still intact that somehow my Magicat would be returned to us.

The next morning my office janitor greeted me with, "Don't you want your cat?" The White House had called again just as he was opening up.

"Where is he?" I was so relieved, I hardly waited for his answer.

"They have him at the northwest gate of the White House, but they say they can't keep him much longer."

We brought Mike home with great rejoicing. He rubbed against our legs and captivated us with his usual charm. He was full of stories of deep-fried treats out of little plastic bags. He'd also developed a taste for pizza. He had hunted worthy prey on vast manicured lawns and in dense shrubbery. I guess telling me the story whet his appetite for a return visit, because three days later he was gone again.

I didn't need psychic talent to tell me the first place I should look was his new home-away-from-home. I phoned the guards at the White House, identified myself, and asked, "Is my cat back?"

"What cat?" came the reply.

It was a different guard! I explained what had happened before and what an exceptional cat Mike was and how important it was to get him back. I could tell the guard liked cats, too. He was sympathetic, but said he had just returned from his vacation and had not met Mike.

But that same afternoon Mike came home, riding like a little king in a White House limousine with a chauffeur and two bodyguards! They had found him strolling majestically across the White House lawn.

The next day Evelyn Peyton Gordon, the then social editor of the Washington *Daily News*, wrote: "Mike is Running for President in His Own Way."

After that Mike was welcome to hunt the White House grounds, but he knew I would be very displeased if he did not get himself home by dark. By then he had become a fine, healthy, and handsome cat with beautiful black-and-white markings. Why would a cat need a pedigree when he was welcome to use the White House for his romps?

After he became famous, producers of television shows regularly called wanting to interview Mike. People sent him clothes and soon he had a large wardrobe. Sometimes he wore farmer's overalls with a big straw hat; sometimes a dinner jacket with a high hat. I never knew a cat would sit still for all that dressing up.

Mike's extrasensory perception matured with him. By the time he was ten years old, Mike and I easily transferred thoughts—and the nice part was, he always agreed with me when it came to advising others.

Mike reserved his most special talents for children. This plump, furry feline somehow sensed what children said to him and, in his own way, he responded to them. I picked up on what he was thinking, so I spoke for him and told his predictions to the children.

Once at a children's party, a beribboned little girl knelt on the floor beside Mike and looked deep into his enormous green eyes. "I want to be an actress," she whispered to Mike. "Do you think I should?"

I told the little girl Mike's prediction: "No, you should be a writer." By the time she was fourteen years old, the girl had several articles published in magazines.

A young boy of twelve years asked Mike to help him choose

between attending school here in the States or abroad. The boy's parents had left the choice to him. Mike's answer, through me, was: "Go to school in the United States. Europe is not advisable—you will have a bad accident!"

The boy shrugged off Mike's warning and went to school abroad. Sure enough, he had a terrible skiing accident.

Like such White House pets as Socks Clinton, Mike the Magicat got tons of mail from children. The volume of letters to Mike became so large that he had his own secretary, Lorene Melton, to help answer his mail.

Several people wanted to do comic strips featuring Mike, and others wanted to manufacture toy cats in his image. Jimmy, my musical genius of a husband, wrote Mike a theme song.

Mike the Magicat lived a long and happy life. Like most Sagittarians, he got along well with people and other animals. He brought many cats home with him, so many that I had a cat pagoda built in my yard where a stray could get a meal on its journey home.

Once, when I had received sad news about my brother, I was alone in my bedroom, weeping. A newly arrived stray padded to the edge of my bed and looked at me, its eyes seeking mine in sympathy. After a while, it came closer and sat there, its eyes saying, "Please don't cry. Everything will be all right."

The cat rubbed its head against me and then curled up next to me. I don't know how long it stayed there. I must have fallen asleep, but when I awoke the cat was still keeping its vigil. It gave me the feeling that my brother would recover. And he did.

Other Remarkable Cats

During his long life, Mike opened the door and welcomed many pets into our home. If Mike didn't have remarkable extrasensory perception, how could he have known that Jimmy had set his heart on a pedigreed cat?

Mike loved Jimmy, but Mike was, more or less, a one-woman cat. So when he was about three years old, Mike "kidnapped" Silver, somehow knowing that the beautiful stray was looking for a special man to love.

When I first saw Silver, it was obvious she'd spent a few nights out on her own, but she was a real lady who had led a pampered existence.

That first morning, in my front yard I saw a slim, silver-brown flash that disappeared before I could get a good look. I knew, however, that it was a cat—and a beauty. When I finally did get a good look at Silver, I knew she was a cat apart from any that I had ever known.

Later the same day, I was ready with a dish of milk on the doorstep. I put it out, closed the door, and waited a half hour before opening it again. Silver was there, ears lifted, poised on three legs as if waiting for an invitation. I called her by the name I'd given her, but she looked right through me with her blue eyes and didn't move a muscle.

I tried calling to her telepathically, but she wouldn't move close enough to let me touch her—even though I sensed she knew exactly what I wanted her to do. I left the milk and through the window saw her return and lap it as soon as I was out of sight. She was hungry, but she held herself worth much more than a saucer of milk.

The next morning, Mike alerted me to Silver's arrival. Through his extrasensory perception, Mike understood that I wanted to get Silver inside, and so he began to herd her in the direction of the door, nudging her toward a plate of food on the doorstep. Mike knew I wanted to feel her body for broken bones or burrs. I also wanted to get her to a veterinarian.

When Mike nudged her she screamed a warning. She seemed to be challenging him to fight, even as he politely got out of her way. Living on her own had improved neither the appearance nor the disposition of this beautiful animal.

I could see the cat had no collar, but it was obvious that she came from a comfortable home. I studied an illustrated cat book and saw that she looked like a fine Siamese. She may even have had a pedigree.

The next morning it became obvious that Mike had a plan. He greeted Silver as soon as she appeared on his well-sprayed garden wall. Instead of menacing her, he paid her some kind of cat tribute. There was no doubt in my mind that he was acting the part of a good host welcoming a guest to his territory.

When Silver walked up to get her morning milk, Mike pounced, landing right behind her with a loud cry. Silver jumped straight ahead, right into our kitchen. Mike had kidnapped her—or catnapped her, if you prefer.

Strangely, once she was inside, Silver made no further protest. I examined her and tried to convey my loving respect. She had no broken bones, and although she needed a bath she was healthy. Silver was peaceful in my hands, but uncommitted.

She surveyed the world around her: tiled kitchen, wood-floored dining room, marble entry—the entire ground floor of our comfortable old townhouse.

Nothing impressed her. Not me, not our house, not even Mike the Magicat. Then Jimmy came into the kitchen, and that was the magic moment. Silver's tail rose proud and straight and she walked right over to the new love of her life. The feeling was mutual.

Instead of marching to the door, Silver regally mounted the stairs and followed my husband to the living room where he'd left his newspaper.

It took me a while to understand that every cat that came through the door wasn't going to be a Magicat. There is no doubt in my mind that Silver understood my messages; she simply chose to ignore me in favor of her true master, Jimmy Dixon. Oh! Unrequited love! I put my heart on a plate and Silver walked right by it.

Later that day, Jimmy took his cat, Silver, to the vet and she came back clean, groomed, and ready to live out her days at his side.

In the weeks that followed, I learned how the Dixons got their one and only pedigreed cat. A lawyer who lived and had his office in a nearby Washington townhouse had died suddenly. Since he was a bachelor, Silver was offered to his nieces and nephews, but there were no takers. Finally, the lawyer's secretary took the cat home with her. Although she knew the woman, Silver ran away—back to the now-empty townhouse where she had been a well-loved queen. Her only thought was to find her lost first love.

I would bet anything that that lawyer wore suits with cuffed trousers; for when Silver approached Jimmy Dixon in his cuffed trousers, she visibly relaxed and began to feel at home.

With surprising ease, Mike accepted Silver as a housemate. Watching these two intelligent cats together taught me that there was more than ESP between them. There were rules as well; rules that were very much like the ones that govern human etiquette.

Silver would let me pet her. Sometimes when only the two of us were in the house she would sit in my lap, but never when Jimmy was available to her—and never when Mike was in the house. In firm cat-to-cat communication, they had staked out their individual masters. How much easier life would be if we humans stuck by the rules of social conduct the way cats do.

Silver was our Scorpio cat drawn to my husband, a Pisces.

One of my favorite true cat stories, comes from Candace Carell who co-created the makeup for the phenomenal musical, *Cats*. Candy's cat, Moritz, actually nudged her toward Broadway and inspired the face of the character of Old Deuteronomy.

Candace, who studied piano at the Julliard School in New York City and was involved with their Opera Theater, showed outstanding talent in all the theater arts. To support herself until she was "discovered," she became a makeup artist, and eventually got a staff job at the NBC television studios. Through the years she collected many Emmy awards for her work.

I met Candy when she worked her magic for the "Late Night with David Letterman" show. I sensed all the talent bound up in this lovely small woman and asked how she chose one career over the others. Candy gives a lot of the credit to her gray tabby, Moritz.

Candy took a leave of absence from NBC in 1979 to follow some promising leads in Hollywood. She said, "I thought I was doing quite well in Hollywood, and it would only be a matter of time before I found my niche. Contacts from the New York theater opened many doors professionally. I'd sublet a flat in Venice, had the company of my cat, and a circle of talented and lovely friends.

"I even thought Moritz was happy, but I was wrong. Something happened to him one day—and he would never tell me what—that turned him off life in California. All I do know is that one day he was missing. Actually, he was mysteriously missing because the flat was completely locked up.

"For hours and hours, friends helped me search. Then, finally, I heard a meow from the basement. There he was, crouched near the window. He jumped into my arms. I asked him if he was all right. He licked me, purred, and told me he wanted to go home—which to him was New York City.

"When I got back to New York, I returned to my television career with NBC and David Letterman. In the years that followed, I had many assignments with the network, including 'Saturday Night Live,' and Tom Snyder's show. I was even Jessica Savitch's makeup artist.

"In 1981, I heard through friends that Andrew Lloyd Webber was bringing his London hit, *Cats*, to Broadway. With the help of costuming and makeup, the American production would turn the musical into a spectacular.

"Who better than me, a woman who talked to cats, to get that job? Every makeup artist in town wanted the job, but I got it.

"My first character makeup design was for the beautiful white cat, Victoria. She was soft and sensual. Slowly, one cat at a time, I created a makeup 'bible,' and as actors were hired, I taught them how to do their own makeup. The musical *Cats* opened in New York in October 1982. Moritz was my inspiration for the face of Old Deuteronomy, the gray tabby cat who was very wise."

Candy's success is not uncommon among those who "talk to animals." Most cat lovers have similar stories to tell about the special extrasensory bond that builds between humans and pets.

My prediction for the future of the relationship between humans and cats is this: There is a cosmic plan for mankind to redevelop extrasensory perception. Someday we will talk to each other with thought transference as well as with language. Pets will play a small but important part in developing our powers of ESP.

Soon, too, science will have some important new information about the relationship between cats and mankind. Already, archeologists have found a cat's jawbone in a caveman dwelling. Was the cat a friend? No other animal remains were found among the human bones, indicating that centuries before the Egyptians "tamed" the cat, this amazing creature was buried in the same manner as humans.

It is no leap of faith for me to believe that the Creator meant for both men and cats to find comfort in each other's company— and so the potential for communication, extrasensory perception, was God-given.

At the Creation we were given the potential for communication through thought transference along with remarkable intelligence. With this intelligence we have developed computers and other labor-saving devices that have opened new worlds of leisure and learning. Our new freedom will be turned to spiritual growth and a striving for world harmony.

CATSCOPES

HOROSCOPES for cats? Certainly! Why not? Cats come into this world under the same stars we do and they live out their lives under the same astral influences as people. No matter what its breed, or mixture of breeds, there is much to be learned about a cat from its birth sign.

When you do not know the exact date of your cat's birth, it may be necessary for you to pick an approximate birth date and the appropriate astrological sign. If you received a kitten for Christmas, for example, and it was between six and eight weeks old at the time, carefully compare your pet's personality with those described for Scorpio (October 24th–November 22nd) and Sagittarius (November 23rd–December 21st), as well as with Libra (September 23rd–October 23rd), just in case the kitten was a bit younger than you assumed when you got it.

Before acquiring a kitten, decide which catscope is most appealing to you and is most compatible with your own birth sign. Visit a whole litter that was born under that sign of the zodiac. Play with and handle all the kittens in the litter. After a few minutes, an inexplicable connection will be made. One of the kittens will reveal itself to be yours.

Experts tell us not to give living things as "surprise" gifts. In the case of cats, the reason is clear; bonding, usually the result of extrasensory perception, can take place quickly, and it is as mysterious as any instance of love at first sight. A child, for example, may be captivated by a cat that looks runty or strange to a parent. The parent, on the other hand, may be immediately drawn to a much "better" specimen. The solution may be to take both kittens home.

Love at first sight is usually the way cats and people get together. When there is an immediate and mutual recognition of an emotional bond then there is a good chance that the cat and the human will forge a deep understanding through extrasensory perception.

A cat that comes to you full grown will soon reveal its probable birth sign if you give it love and encourage it to talk to you. Bonding with an older animal can take a little longer than with a kitten, but it can and does occur at all ages.

Always remember that your pet's future is not just in the stars. It is also in your loving hands.

THE ARIES CAT

(MARCH 21ST–APRIL 20TH)

If there ever was a party animal it is the Aries cat. When a stranger comes to the door it's "Let me at him!" "He's mine!" "He loves me!" Underneath the brash exterior, however, there is a vulnerable core seeking affection. Only after you declare your love will the Aries cat decide whether or not it likes you. As quickly as it rushed you when you met, the Aries cat could turn its back on you, or assume that you are friends for life.

Cats born under the sign of the astral Ram have the same personality traits as Aries people: They are headstrong and impulsive. Like a ram, an Aries cat is prone to charge head-long into a situation. When an Aries cat wants to run across the neighbor's lawn, it will trample the "Keep Off the Grass" sign. When it hears the bell of the ice-cream wagon, it might jump over the fence to chase after its favorite deli-cacy. The wise owner of an Aries kitten will begin early to direct this enthusiasm into the right channels.

Although Aries is unlikely to become a soulful lap cat, it has no lack of its own brand of extrasensory perception. Aries has a kind of clairvoyant understanding of cause and effect. It will know when you need to laugh and how to make you do it. It will also sense when you are hungry and lead you to the refrigerator, where it is likely you will reward it with a snack of its own. (You may, incidentally, have to watch this cat's weight as it matures.)

Like cats born under the sign of Leo, the Aries cat is a natural leader. In a feline argument, Aries is the likely winner. When it is time for an outing, an Aries cat will rush you to the door and show you where it is—just in case you have forgotten.

Your Aries pet is tenacious. When it feels it is right, it will not give up easily. That is the reason you must learn to respect your cat's rights, just as you want it to respect yours.

The Aries cat will get along with other pets in the household if it is the acknowledged leader. For the most successful results in training, it is best to teach an Aries cat the rules of your household when it is a kitten. This doesn't mean it won't do as it pleases, but it will want to please you.

It is important not to let anyone annoy an Aries cat while it is eating. And when it feels like a nap do not disturb it. When the young Aries is not resting it tends to be a whirlwind, and it loves the outdoors. Like an Aries person, your pet does not demand much; but the few things it insists upon are essential to its well-being.

People who get along best with Aries cats are those born under the signs of Leo, Virgo, Libra, and Scorpio.

The Taurus Cat

(APRIL 21ST–MAY 21ST)

Taurus is a nice cat to be around. Ruled by the planet Venus, the Taurus cat is usually lovely to look at, and loving to be with as well. It can be the most contented, home-loving, and loyal of pets.

As a kitten, Taurus is a whirlwind. Children love this cat's exuberant energy. The young Taurean's natural domain is outdoors, where its hunting skills and agility are well tested. Indoors it will explore every nook and cranny, looking for the most precarious spot on which to catnap. Boisterous Taurus brings its own sunshine into the lives of its human family.

It is in times of trouble that a Taurus cat shows the depth of its extrasensory perception. Sensing when its master is worried, the cat will offer to share its playthings. The Taurus cat's exuberant love of life can't help but make everyone in its family smile.

Cats born under the sign of Taurus share the qualities

of the astral Bull: strength, courage, endurance, determination, pride, and enthusiasm. What a powerful personality!

This combination of traits ensures that this is not a cat to back off from a fight, and as a kitten it may swashbuckle its way into some dangerous situations. Inborn bravery makes a Taurus cat a daredevil, and it sometimes take risks too lightly. Occasionally, its reach goes beyond its grasp. Discourage your Taurus kitten, if you can, from leaping from high places or from challenging big dogs. The positive side of Taurean bravery is that they make good "watchcats." A Taurus cat will protect home and master with its life.

If encouraged to breed, Taurean queens are the most loving and attentive of mothers, and if a stranger threatens they race into battle.

Sometimes your Taurus cat will ignore you for days or even leave home for a short period if you interfere too insistently with its plans. Despite the Taurean reserve in matters of love, this cat has deep feelings of loyalty and is possessive of its master.

Let the Taurus cat tell you how much affection and attention it wants and when. Since the cat believes it has you completely "sized-up," it expects you to understand its need for love as well.

Begin training a Taurus cat as soon as you bring it into your home. Taureans like the comfort of orderly, predictable routines. They are quick to recognize patterns, and slow to change them. Your Taurus cat will be unhappy to find its food dish in the wrong place, and will push it back if you move it. If changes have to be made, an understanding master will reassure a confused pet with lively conversation, warm smiles, and silent messages that tap into the cat's extrasensory perception.

An adult Taurus does not need to be top cat, but it demands respect. After it becomes accustomed to other pets in the home, it is usually happy to share.

A Taurus cat tends to overeat so be careful about feeding. It likes the feeling of fullness in all things—and will hang around the table looking for a share of your dinner after it has finished its own.

People who get along best with Taurus cats are those born under the signs of Capricorn, Sagittarius, Cancer, and Gemini.

The Gemini Cat

(MAY 22ND–JUNE 21ST)

The Gemini cat is full of surprises. One day it may rush to you at the door. The next day it may saunter into the room minutes after you call it. Nothing is wrong. You just have two cats masquerading as one! But both sides of the Gemini cat's personality are loving, intelligent, and faithful. And both sides have highly developed extrasensory perception.

Gemini is the Latin word for "twins," and those born under this sign—pets and people alike—reflect the two-sided nature of its astral symbol. Their unusual ESP makes them extremely understanding of changing human moods. They are slow to anger and quick to forgive—they sense that you can have an occasional bad day.

They are great talkers with very definite opinions, so your Gemini cat may meow a bit more than most other cats, especially if it's a Siamese.

A Gemini kitten is quick, fun-loving, and busy. In the

flick of a whisker, it can go from a sound sleep to full alert and is always ready for action. Children love Geminis because they are sweetly adaptable, and are ready to get involved in any new game.

Although any kitten is best trained early, a Gemini will determinedly learn the rules of the house, no matter what its age when it comes into your life. Gemini cats seem to be able to learn new tricks at any age. Adaptability also ensures peaceful moving days and great vacations with your cat.

Of all cats, Geminis take best to surprises and unexpected changes. Now and then they enjoy something unusual for dinner and often like to sleep in different rooms on different nights.

A Gemini cat is so fascinating, even to itself, that it can be left alone all day without becoming too lonely or bored. But Gemini can be a worrier, so if you are late, or go off for a vacation, you come home to find a rather sulky cat demanding to know where you have been.

Although intelligence and curiosity make Gemini cats delightful companions, there is another side to these personality traits. Their inquisitive nature can make them nosy. This can be particularly annoying when your Gemini cat insists upon inspecting everything in the house or in a neighbor's yard.

Gemini's adaptability can give the impression that the cat is a bit fickle with its affections. Yesterday's favorite toy is ignored today. A taste for new experiences and new environments may even lead the cat to run away from home, and when you track it down a few blocks away, enjoying the hospitality of the new family in the community, it will not be the least bit repentant. This cat just likes to meet new folks!

A Gemini cat is as enjoyable as a three-ring circus. We all have our different roles to play in life, and the role of this cat is to enliven the world with its frisky personality and delightful ways.

People who get along best with Gemini cats are those born under the signs of Sagittarius, Aquarius, Leo, and Capricorn.

The Cancerian Cat

(June 22nd–July 22nd)

The cat born under the sign of Cancer, the Crab, will always be at the door to welcome you home. Once you win this cat's love, it never wavers. The moon, ruler of those born under the sign of Cancer, makes affection particularly important to this little home-lover.

Cancerian cats are usually so gentle and sweet that they are favorites of children and elderly people. They adapt best to small households because they love peace and quiet.

What is true of the female, however, does not apply to the young unneutered male. He may not be the first out the door in mating season, nor will he get in catfights if they can be avoided, but he has so much charm that females find him.

The charm of a cat born under the influence of the moon is the result of a particular extrasensory perception that makes it adept at gauging and matching feelings, especially those of the people it loves. Empathy is the Cancerian cat's middle name.

Cancerian ESP gives these cats the ability to anticipate people's moods and actions. Just think about putting a Cancerian cat outside on a cold morning and it will hide under the bed until you forget the idea. They love to play games, especially hide and seek, and are content to be alone in the house for hours.

The Cancerian cat is quick to inspect every life form that comes near your home—and to make an instant judgment about whether the person, animal, other pets, or delivery is acceptable. This can be very disturbing if the mailman hasn't made friends with your protector, but it also makes the Cancerian cat a very good "guard cat."

Since a young girl often grows up closely bonded to a moon cat, when she begins dating and anticipates marriage it can be a bit of a problem. Any young man who comes through the door gets a thorough inspection. Remember that the cat is an excellent judge of character!

Of course, the Cancerian cat also comes to regard its home as its castle. It wants the highest perch; and it isn't especially fond of sharing space with another cat, unless they were raised together as kittens.

The heavenly sign for Cancer is the Crab, and sometimes your perfect housemate will show a little "crabbiness." People as well as animals who were born under the influence of the changing moon can be a bit moody.

Female Cancerian cats, and older male Cancerians, are true lap cats, willing to snuggle for hours at a time. They are never happier than when your hand is ruffling their fur. And they give back oceans of affection.

The Cancerian cat is extremely cautious. That is why your pet is often reluctant to explore new walkways and hiding places.

The Cancerian cat might disappear for hours—and although it is somewhere in the house, or very close to home, you will never find it until it wants to be found. Cancerian pets, like people born under this sign, need time alone to collect their energies for responsibilities, which they take very seriously. Give your cat the space it needs and it will soon come bounding back to find you!

People who get along best with Cancerian cats are those born under the signs of Pisces, Taurus, Scorpio, and Aquarius.

The Leo Cat

(JULY 23RD–AUGUST 23RD)

Leo cats are ruled by the sun, the most vital life force. Their motto is: "Live life to the fullest—enjoy it, consume it!" And to give this cat even more self-confidence, the Lion, the king of the beasts, is the symbol of this sign.

It is amazing how accurately ancient peoples associated signs with animals. The Leo cat, for example, is uncanny in the way it imitates the lion; it is proud, loyal, and can be fierce.

When Leo the cat uses its extrasensory perception, it is often for its own benefit. Thinking of a snack? Guess who you meet by coincidence at the refrigerator? Want a snuggle? This rather self-centered cat will ignore you as long as it can. But when you have its attention, a Leo cat is warm and generous. It senses that it is a star, and wants to dazzle you with its love when it welcomes you home.

While the mature Leo cat marches about in a slow and stately fashion, the youngster is capricious. It is also impulsive and powerful. Even a little Leo kitten is stronger than you would expect and is always dashing off as if it just thought of something that must be attended to immediately.

Leo enjoys its beauty. Of all cats, Leos seem to actually like being groomed. They appear to realize it helps them look as regal as they know themselves to be. Leos are theatrical. They love to show off. This is their way of asking you for a compliment and a pat on the head. Perhaps you'd like a quiet game of ball rolling—the Leo is liable to take you on a merry chase. It is always overdoing things.

Like the noble lion, the Leo cat is a born leader, always in front of the pride and sometimes endangering itself with its recklessness. The Leo cat will fight for its territory. If it is to share the house with another pet it should be one much older or younger. This is a top cat we are talking about!

Once its dominance in the household is secure, the Leo cat becomes tolerant—and only after the exchange of much affection and bonding does it reveal the nurturing side of its nature. Most of the time, "Live and let live" is the creed of the Leo cat. Consequently, it won't go out of its way to find trouble.

The secret the Leo cat doesn't want everyone to know is that it is shy. But once this cat is your friend, you have a friend for life.

People who get along best with Leo cats are those born under the signs of Gemini, Libra, Aries, and Pisces.

The Virgo Cat

(AUGUST 24TH–SEPTEMBER 22ND)

The Virgo cat is as mysterious and as secretive as a sphinx. No matter how much this cat trusts you, it will have a secret life. It may hide its toys or bury food, or have a private place under the house where it can meditate alone. It may be that Virgo's mystic extrasensory perception reveals a world beyond what our eyes can see.

This unique ESP helps Virgo observe and understand things overlooked by other pets. And while all cats have sensitive noses, the Virgo cat seems to have especially delicate senses of smell and taste.

If you pay attention to your Virgo cat, it could save you from catastrophes large and small. This cat will search out sources of minor odors which no one else has noticed. Many of these sensitive creatures have saved their masters' lives by warning of fires or gas leaks at the first scent of trouble.

While the Virgo cat uses its extrasensory perception

every day for its own advantage, this cat has enough to spare for you and your whole family. The Virgo cat knows when you are ill or unhappy. At those times it will be especially quiet, remaining near you to listen to your troubles and lick your ear.

Not only is Virgo sensitive to smell, but it is also sensitive to sound and light. This is the artist among cats. This talent is evident in the way it keeps itself and its habitat as neat and as beautiful as possible.

The Virgo cat is devoted and faithful, aptly associated with its star sign, the Virgin. After this cat has chosen you as its master, it may not accept affection from anyone else.

Although most of the time Virgo seems like a loner, this cat is attuned to everything around it. It especially appreciates order, cleanliness, and people who are very predictable.

Like Virgo people, Virgo pets demand a clean and healthy environment. Your Virgo will be happiest right after a grooming and will sulk if its dish and litterbox are not regularly cleaned.

You will know you have a Virgo's complete love when it grooms you with tender flicks of its tongue while nestling on your shoulder.

While it may seem sweet and shy, Virgo is delightfully adventurous. However, the cat who is happiest lying in the sunshine near a patch of green grass is probably a Virgo. Your Virgo pet may sometimes sniff the flowers in your garden or lie quietly at your feet when you listen to music. As the cat ages, you will have to watch its diet, for Virgo is a master of conserving energy—especially as it matures.

Ruled by the planet Mercury, which is depicted by the god with winged feet, Virgo is likely to dance away from danger. This is one of the most peace-loving of cats.

People who get along best with Virgo cats are those born under the signs of Libra, Capricorn, Aries, and Scorpio.

The Libra Cat

The eyes of a Libra cat reveal remarkable extrasensory perception. A Libra is the cat most likely to look you directly in the eye, as if searching for the slightest suggestion of encouragement or disappointment—and it never wants to disappoint. Of all cats, the Libran may have the most ESP. It seems to be able to communicate with other cats, as well as with humans, with hardly a flick of the tail.

If you desire a companion pet that is almost human in its understanding, Libra may be the cat for you. Ruled by the planet Venus, the Libra cat gives back as much love and devotion as it gets.

The young Libra is a dancer. You will recognize this cat by its grace and sense of balance in all things. Libra's astral symbol, the Scales, indicates how Venus enhances its remarkable extrasensory perception. On the scales of love, this cat will try to match you gold piece for gold piece.

The Libra cat responds well to the most affectionate

people. It will actually gauge its responses to yours. If you are lavish with praise, grooming, and love, the cat will shower you with affection. But if you are in a bad mood and shoo it away, expect to see yourself reflected in a very sulky cat.

Libra's sense of balance makes for an excellent housemate. This cat is not likely to knock over household objects or become hyperactive. And as the Libra cat grows older and wiser, it has a tendency to be a bit languid.

This desire to take life at a slow pace can affect Libra's weight, so you must be sure to give it slightly less food than your more active pets as it grows older. The young Libra usually self-regulates its diet so that it is perfectly balanced —neither too skinny nor too fat.

The Libra cat enjoys its good looks as much as you do, and grooming itself is one of life's chief delights.

Just like Libra people, the Libra cat is curious and inquisitive. It is always off exploring some corner of the house or garden or nosing around underneath tables and in closets. It loves to watch other animals at play, but rarely picks quarrels with them. This cat avoids conflict with people, too. It doesn't want to annoy its master.

Harmony at home is very important to the Libra cat, and it will run and hide if your household is in chaos. If raised among boisterous children, the Libra cat will adjust well, becoming an island of tranquillity.

The Libra cat asks for little from its master: a balanced diet, an even-handed disposition, and a peaceful home will earn you much in return.

People who get along best with Libra cats are those born under the signs of Taurus, Aries, Aquarius, and Gemini.

The Scorpio Cat

(OCTOBER 24TH–NOVEMBER 22ND)

In the constellation of Scorpio are some of the brightest stars in the heavens. And, here on Earth, Scorpio seems to be the dominant sign of movie stars and other celebrities.

Scorpio is the sign of intense and passionate people and pets. The symbol of this sign is the Scorpion, which is said to be dangerous when attacked. The extrasensory perception of the Scorpio cat seems to be honed toward one specific human, and to a mystical interaction with the forces of nature.

If all the stories of homing cats were documented, a large proportion would likely be Scorpios. A Scorpio cat is the most devoted of companions. This is a cat who will never leave you.

This does not mean you won't find a Scorpio missing at times—you will. This cat can be secretive and self-contained, but rest assured, you are a magnet that will draw it back to you as long as there is strength in its body.

When interacting with humans, the Scorpio's use of extrasensory perception is a bit different than other cats. While it is getting acquainted, the Scorpio cat will put you through a series of tests. It is always watching and will know all your secrets, but rest assured that Scorpio is one creature who will love you without reservation.

When visiting a litter, you will find the Scorpio kitten to be playful and charming. But the kitten you first select may not be the kitten you take home.

Winning the affection of a Scorpio is harder than with some other cats. In matters of love, it is no pushover. The Scorpio cat won't take to everyone, and there is no use hoping that its feelings will change.

The Scorpio cat has one master at a time, and once a bond is established, the Scorpio cat worships its master for life. The Scorpio cat is a Princess or Prince Charming; its heart is full of passion.

The first time you hug your Scorpio kitten, you will notice its forceful spirit. When happy, this cat will throw the whole house into a jumble of contagious joy. But when it is sad or lonely, it can break your heart with its crying.

Scorpio is the sex symbol of the zodiac, and Scorpio cats are vital, willful, sensual, and beautiful. This cat can be a saint or a sinner, and may be both.

Although Scorpio cats tend to be quite slender and sleek when they are young, they are deceptively strong and robust. Because of the male's athletic nature, you should make an early decision about whether you want your animal neutered.

This intensity of personality suggests that a Scorpio kitten will accommodate itself better to a new environment than other cats. It loves life and will want to travel with you, even if it is kept inside most of the time as a house pet.

Rugged and adaptable, a Scorpio cat may look small and scrawny, but it has nerves of steel and immense determination. It will keep trying to jump a troublesome fence until it clears it, and will pester you for a second helping of dinner until you give in. Keep an eye on the Scorpio's weight as it ages. And be sure to give this cat the extra love it takes to satisfy its powerful craving for affection.

People who get along best with Scorpio cats are those born under the signs of Cancer, Capricorn, Pisces, and Taurus.

The Sagittarius Cat

(NOVEMBER 23RD–DECEMBER 21ST)

A Sagittarian cat is as confident as the starry Centaur—which is its symbol in the zodiac. This is a cat who loves a challenge and is dependable in good times and bad.

A Sagittarian is a romantic cat. It will woo you with little gifts and enjoy long sessions of petting. On moonlit nights, young males like nothing better than to serenade their nearby ladyloves with howling melodies.

Sagittarians are also great actors. If the Sagittarian cat doesn't actually have finely developed extrasensory perception, it will make you think it does. It gives affection in robust quantity. This is the cat that sleeps at your feet, bounding onto the bed as soon as you are settled. The Sagittarian is a wonderful companion—especially for frolicking children.

Its ESP allows the Sagittarian cat to see the big picture. It knows who its friends are and is usually frank and candid with its opinions. If a Sagittarian doesn't like its food or living quarters, you'll know it. This cat is not known for its cunning, so if it makes a mistake, it will stand by it—waiting for you to say you love it anyway.

Sagittarians are happiest as outdoor cats. They love to hunt, and stalk through the flower garden with the determination of a bloodhound on the trail of an escaped convict. If a Sagittarian cat is kept indoors, it should be given ample space to roam.

One of the great charms of the Sagittarian cat is its unpredictability. This cat is delightfully impulsive. Without meaning to disobey, the Sagittarian will ignore your calls when in pursuit of a rabbit or a bird.

Until trained, the impetuous, darling Sagittarian kitten is a hazard in a household of breakable treasures. But Sagittarians are such wonderful housemates that most people delight in their antics.

Although the Sagittarian kitten may be hard on the crockery, the Sagittarian cat is one of the easiest to fit into households with other pets—even other cats. It is easygoing and relaxed in relationships, and if another cat becomes a favorite, the Sagittarian will back off and find another person in the household to lavish with its loving attention.

The Sagittarian likes bigness—and will become as big as it can. This healthy appetite can lead to a very large and stately cat in old age.

But no matter how it matures, the Sagittarian is the free spirit of zodiac. It is unconventional, and will try anything once. Although it isn't interested in repeating tricks, it might perform a favorite stunt again, just for you, on a day when it knows you need a big laugh.

If you have won the love of a Sagittarian cat, you are blessed. This is a self-sufficient animal who will always be proud, and will treat you as an equal. If there is such a thing as a man's cat, it may be the Sagittarian of either sex. This is a cat that appreciates praise, but is usually happier with a manly pat on the back than with an evening of cuddling by the fire.

People who get along best with Sagittarius cats are those born under the signs of Aries, Aquarius, Gemini, and Cancer.

The Capricorn Cat

(DECEMBER 22ND–JANUARY 20TH)

If you find a beguiling kitten who thinks it's human, it is probably a proud and intelligent Capricorn cat.

Capricorn cats have a sweet nature and very highly tuned extrasensory perception. They seem to pick up the smallest change in human feelings. They may not always be playful when you are, but they will surely fly to your side in moments of sadness. Like human friends, they feel they owe support to those they love.

The tiny Capricorn kitten doesn't always look like the pick of the litter, but it grows up in the most beguiling way, changing from frail to robust and beautiful by the time it becomes master of the house.

The Capricorn adapts well to a life indoors. Its unusual extrasensory perception makes it as dependable as a guard dog. This is the cat who will warn you if someone emits bad vibrations.

Alert and inquisitive, nothing escapes the Capricorn cat. Even when napping, its mind is busy and its half-closed eyes are keeping track of everything happening in the household.

Finely tuned ESP and intelligence makes a Capricorn cat easy to train, especially if you lavish praise on its talent for quick response. A Capricorn loves the limelight.

A Capricorn cat is meticulous in its grooming, and you as its master will get a hint of the depth of its love when a helpful paw attends to your grooming.

Like their zodiac symbol, the starry Goat, Capricorns are the most agile and fleet-footed of cats. Unlike most other animals, your Capricorn cat will deftly walk around a puddle rather than splash through it.

Because the Capricorn cat is more serious than others, it may seem haughty and aloof. This is just a sign of independence. The greatest token of its affection is to treat those it loves as equals.

Don't be surprised if your Capricorn seems to change its personality when let out of the house. It has a strong hunting instinct and an adventurous spirit.

Capricorns are famously long-lived creatures. That makes them good companions for small households where pets are often loved like children.

A Capricorn cat loves order and routine. It will find a sunny windowsill and claim it daily for years—always finding new delights in the play of the light, or the warmth on its fur.

Capricorns like to be in charge of their own time. Play

with your Capricorn cat when it wants to play—and be ready for some very serious nestling when it is in the mood.

A Capricorn cat needs respect as well as affection. It is nobody's fool and does not like to be teased.

People who get along best with Capricorn cats are those born under the signs of Aquarius, Pisces, Virgo, and Leo.

The Aquarius Cat

(JANUARY 21ST–FEBRUARY 18TH)

Don't consider an Aquarian cat unless you are ready to make a serious, lifelong commitment. You will have a friend for life. If it is abandoned, the Aquarian cat can perish from a broken heart.

Extrasensory perception between the Aquarian cat and its master builds slowly at first, but deepens as each year passes. This cat may not show friendship by exuberant exhibitions, but, like an intellectual friend, it will match you step by step as you go about your daily business. Aquarians often turn out to be geniuses, or, at least, cats with extraordinary brain power.

The uncanny intuition of an Aquarian cat makes it seem almost human. With remarkable extrasensory perception, it will react to conversations as if it understands what is being said. Aquarians seem especially attuned to the changing moods of children. And as long as a child isn't cruel, this cat will be a teaching companion as well as a playmate. It will allow itself to be carried around in the arms of a child and then cuddle up with its young companion for a nap.

As you may have noticed, Aquarius is the sign that gets along with all kinds of people and other pets. But this is not the kind of cat who wears its heart on its sleeve. At first it is sweetly shy with strangers.

When you have a party, your Aquarius cat will retreat until the house is back to normal. This is not because it is high-strung; it simply knows who its friends are and prefers to save its energy for the family.

If you are looking for a cat who can fit in with a multigenerational household, find yourself an Aquarian.

Usually—almost always—is the modus operandi of the Aquarian, whether human or feline. Influenced by the planet Uranus, Aquarious is the most unpredictable sign of the zodiac. The Aquarian cat won't allow itself to be taken for granted. Eventually it will do something that will make you shriek with laughter or amazement.

There's an unconventional streak in all Aquarians. Your pet may have a taste for exotic foods or it may have a favorite sitting place—under the table or in the middle of a tulip bed—right where you don't want it to be. That's the free spirit coming through.

Like the Waterbearer, its astral symbol, Aquarius pours out surprises. Some Aquarian cats learn to open doors, turn television sets on or off, get water out of a pump or faucet. They do things you would never expect cats to manage.

An Aquarian is as good on a farm as in a city apartment. This cat is usually brave—especially when its family or its home is threatened.

An Aquarian cat can be loving and gentle, unconventional and sensitive, changeable and shy. But underneath its quixotic façade beats the heart of a loyal companion and a worthy friend.

People who get along best with Aquarian cats are those born under the signs of Virgo, Gemini, Leo, and Libra.

The Pisces Cat

(FEBRUARY 19TH–MARCH 20TH)

For cats as for people, the symbol of Pisces is a pair of fish swimming in opposite directions. That's why a Pisces seems to go two ways at once.

The most common way for the lovable Piscean cat to show its nature is to be faithful and loving six days a week, and then take the seventh day off to roam at will. Never fear, this cat will come back. The Piscean's special blend of extrasensory perception makes it a homing cat. If left for the night in the basement, it will strain every muscle in its body to climb upstairs to its master's bed.

You will never have a sweeter companion than a Piscean cat. Its extrasensory perception is especially attuned to sadness. The Piscean cat will stay near you until it lifts your spirits. Pisces is romantic, loyal, and will make itself an island of tranquillity to calm you after a day of activity and stress.

The Pisces is a cat you should acquire as a kitten, so

that you can easily teach it the house rules. At a young age, this cat is an extremely quick learner, which means that it can pick up bad habits as well as good ones.

With the house rules out of the way early, Pisces is free to divide its time between investigating the physical world as far as it can travel and its extrasensory life where no human can follow.

With so much going on in its world that is beyond human understanding, you can't expect an instant response to your calls. It takes time for Pisces (the double fish) to change from one direction to another. Add to that the fact that it loves tricks—choose a Piscean cat only if you thrive on surprises.

Pisces is very good at heart, loving, and eager to please. Once taught, it usually does the right thing—unless it wants to get your attention or create some excitement.

Pisceans are born under the influence of the planet Neptune and they all seem to be explorers and travelers. Many wandering cats are Pisces. But the same trait that makes the Pisces cat a roamer also makes it a delightful travel companion.

Be warned that a Pisces kitten doesn't know its own vulnerability. Its desires are sometimes beyond its reach. Pisces is apt to be too confident, believing it can ramble around the neighborhood without mishaps. Keep a close eye on it. The owner of a Pisces kitten must be careful not to let it over-exert itself.

The home-loving side of Pisces makes sleeping in a sunny window a sensual experience—and it will enjoy its food to its own detriment if you give in to its charming entreaties.

Neptune gives Pisces an imaginative brand of extrasensory perception. You will often see it anticipating what comes next. This makes Pisces a cat who is confident of your return if you leave it alone all day while you work. While you are gone, you cat will fill its time playing with imaginary playmates and fighting imagined enemies.

People who get along best with Pisces cats are those born under the signs of Scorpio, Libra, Cancer, and Virgo.